The Human Body

Disney PRESS

New York

The Human Body
is produced by becker&mayer!
Bellevue, Washington
www.beckermayer.com

For more information address Disney Press, 114 Fifth Avenue, New York, New York
10011-5690

Printed, assembled, and manufactured in China.

ISBN: 1-932855-59-9

Library of Congress Cataloging-in-Publication Data on file.

First Edition

10 9 8 7 6 5 4 3 2 1

06345

Visit www.disneybooks.com.

Written by Paul Beck
Edited by Ben Grossblatt
Art direction and design by Andrew Hess
Design assistance by Claudia Vellandi and Scot Burns
Biology illustrations by Jennifer Fairman
Kid illustrations by Jim Bradrick
Production management by Jennifer Marx
Facts checked by Melody Moss
Special thanks to Ryan Hobson

Head to Toe

Everything you do—moving, eating, breathing,
talking, and even thinking—you do with your body.
Hundreds of bones and muscles and *billions* of cells
work together to make these things happen.
Even with everything that's going on inside, your
body works better than the best machine ever built!

No two bodies are exactly the same. The color of your
eyes and skin, the speed at which you grow, how tall
you are, the shape of your bones, and everything
else about your body adds up to make you a unique
person, different from everyone else. At the same time,
everybody's body has the same parts, and the parts all
work the same way.

In this book, you'll get a look inside your body. You'll
explore what your body is made of, how it is put
together, and how all the parts work together to make
one amazing human—you!

YOU'RE AMAZING,
HEAD TO TOE!

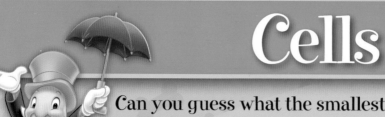

Cells

Can you guess what the smallest parts of your body are? They're smaller than your tiniest eyelash, even smaller than grains of sand. They're cells. Cells are really, really tiny. Take a look at the tip of your finger. You're looking at millions of skin cells!

Different cells, different jobs

Some cells make things, some cells move things, and some cells are the building blocks that form the parts of your body.

Red blood cells carry oxygen and nutrients to your other cells. It takes just 20 seconds for a red blood cell to circle through the whole body!

Skin cells keep your insides in and your outsides out. Your skin is only about as deep as the tip of a ball-point pen!

Muscle cells move the parts of your body. Did you know it takes 72 muscles to speak a single word?

Some bone cells help build bones. Some help dissolve them. It's all part of bone growth.

THEY ALL WORK TO MAKE YOU WORK!

Nerve cells (neurons) carry messages between your brain and the rest of your body. The average human brain has about 100 billion nerve cells!

Cell

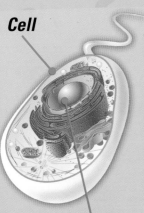

Nucleus

Dividing and multiplying

Your body makes new cells all the time. For example, your body makes and destroys about eight million blood cells every second! First, each cell copies the DNA inside its nucleus. Then the cell splits in half to make two cells. Each new cell has a copy of the first cell's DNA.

THERE'S ONLY ONE YOU, AND THERE'S ONLY ONE ME!

Unique you

There's a copy of the same DNA instructions in every one of your cells.

What's even more amazing is that every copy of your DNA also has the instructions for building a whole human being—you! Unless you're an identical twin, no one in the world has the exact same DNA pattern as you.

DNA molecule

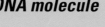

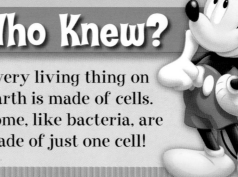

Who Knew?

Every living thing on Earth is made of cells. Some, like bacteria, are made of just one cell!

Your Skeleton

Cells make up all kinds of things in your body, including bones. And where would you be without your bones? Not walking around, that's for sure. You'd probably be on the ground—just a big blob! Bones make up your skeleton, which holds your body up.

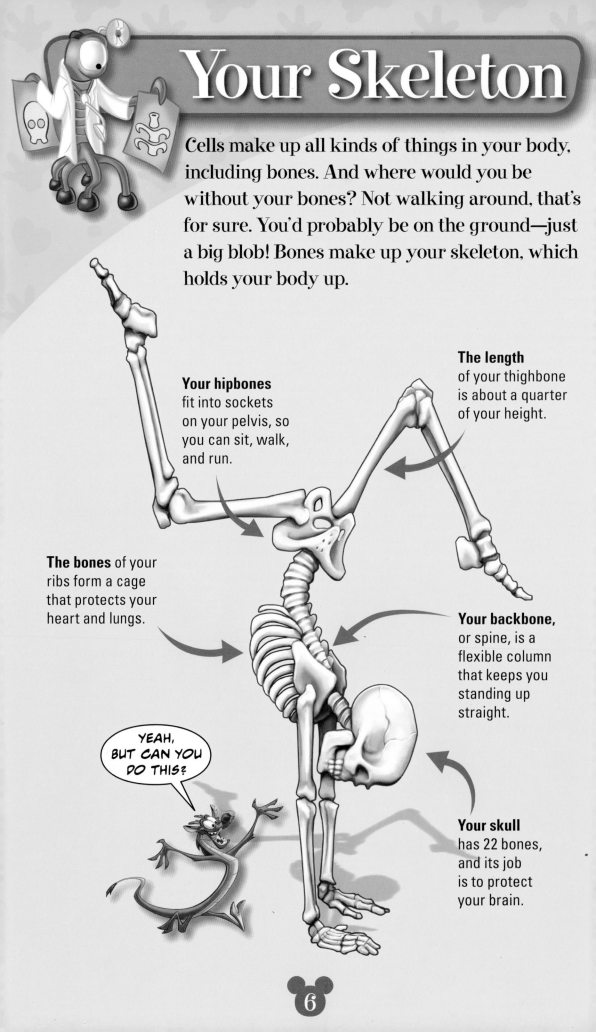

Your hipbones fit into sockets on your pelvis, so you can sit, walk, and run.

The length of your thighbone is about a quarter of your height.

The bones of your ribs form a cage that protects your heart and lungs.

Your backbone, or spine, is a flexible column that keeps you standing up straight.

YEAH, BUT CAN YOU DO THIS?

Your skull has 22 bones, and its job is to protect your brain.

Bones, outside and inside

The outside layer of a bone is hard and solid. It's called **compact bone**. Inside the compact bone layer is a material called **spongy bone**. (It's called spongy bone because it's full of holes, like a sponge, not because it's soft.) Spongy bone is strong but light.

THAT'S WHAT I CALL A BIG BONE!

Compact bone

Spongy bone

They're alive!

The bones you see in a skeleton at a museum are dry and brittle, but the bones inside your body are alive. Blood vessels run through your bones. Nerve cells inside give your bones feeling.

BARE-BONES FACTS

- The strongest, hardest stuff in your body isn't bone—it's tooth enamel.
- Bone is stronger than concrete.
- The longest, heaviest bone in your body is your *femur* (FEE-mer), or thighbone.
- The smallest, lightest bones in your body are deep inside your ears. The **stirrup** bone is the smallest one.

Stirrup bone, actual size!

Bones

Bones that grow are rubbery bones. When you were born, your bones were a lot more flexible than they are now. Much of your skeleton was made of tough, rubbery stuff called *cartilage* (KAR-tuh-lij). You can feel what cartilage is like if you squeeze your nose or the stiffer parts of your ears (not the lobes).

Big change

As you grow, most of the cartilage in your skeleton gets turned into hard bone.

- The long bones of your arms and legs have growth plates inside them. These are always making more cells and making the bone longer. As you get older, the growth plates get replaced with hard bone.

- As the bones get longer, you get taller! Your bones will be mostly finished growing when you are 18 or 19 years old.

Growth plates

TOO BAD YOU CAN'T GROW INTO A BEAUTIFUL BUTTERFLY!

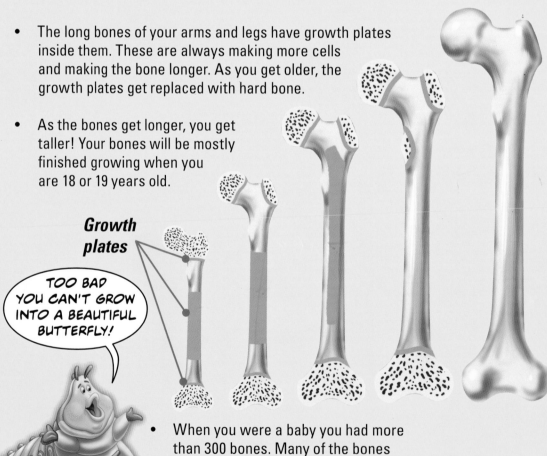

- When you were a baby you had more than 300 bones. Many of the bones joined together as they grew. Now you have about 206 bones.

Heads up!

Your skull isn't just one bone, it's 22 separate ones.

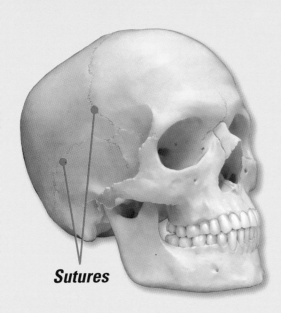

Sutures

- Your *cranial* (CRAY-nee-ul) bones form the round part of your skull that holds and protects your brain. There are eight flat bones. They are fused together at joints called **sutures** (SOO-chers), which look like squiggly lines.

- Your face has 14 separate bones. They form your eye sockets, the roof of your mouth, your cheekbones, and your jaws.

Watch your head!

Your bones are strong, but if you hit your head hard, you can bang your brain into the inside of your skull or even break your skull bones. That's why smart kids protect their heads with helmets when they ride bikes, scooters, and skateboards, or do other activities where there is a risk of head injury.

ONE PROTECTED HEAD, COMING THROUGH!

SPINE-TINGLING

Let your chin hang down on your chest and touch the back of your neck. Can you feel those bumps that run down your back? Those are bones called **vertebrae** (VER-tuh-bray), and they make up your spine. Okay! Stand up straight again! Time to stop slouching!

Joints

Now that you know all about your amazing bones, let's take a look at how they help you move! To get around, you need not only bones, but also joints—the places where your bones meet!

3 ways to move

Your skeleton has different types of joints that move in different ways. Check out these three kinds of joints on your very own body!

Ball-and-socket
Swing your whole arm around in a circle. You've found a ball-and-socket joint. A **ball-and-socket joint** has a round part (the ball) that slides around in a cuplike part (the socket). These joints move in all directions.

Hinge
Now bend your arm at the elbow and straighten it again. That's a hinge joint. A **hinge joint** bends and straightens like a door hinge.

Pivot
Finally, shake your head as if you're saying "no." You're using a **pivot joint**. "Pivot" means "turn in one place." A pivot joint is a twisting joint.

HIPS AND SHOULDERS ARE BALL-AND-SOCKET JOINTS!

Ball-and-socket

ELBOWS AND KNEES ARE HINGE JOINTS!

Hinge

THE JOINT WHERE YOUR SKULL MEETS YOUR SPINE IS A PIVOT JOINT!

Pivot

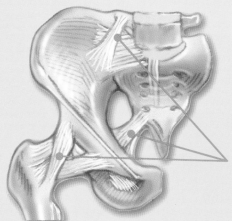

Keep it together

Ligaments are bands or sheets of strong fibers, like strong rubber bands. They hold bones in place.

Ligaments

Do double-jointed people really have double joints?

Some people have joints that bend in directions most people's joints won't go. Those people are often called "double-jointed." That doesn't mean they have two joints where most people have one. It just means the joint is extra flexible.

WE'VE GOT ONE FLEXIBLE KID HERE!

LONELY BONE

Every bone in your body connects to another bone, except one. Your *hyoid* (HI-oid) bone, at the top of your neck just below your jaw, is all by itself. You can feel your own hyoid bone this way:

1. Very gently put your thumb and first two fingers on either side of the front part of your neck, just below your jaw. Don't press hard!
2. Now swallow. You'll feel your hyoid bone jump up between your thumb and fingers.

Busted!

Your bones are tough, but they're not unbreakable. Imagine a pencil—if you hit it hard enough or bend it too far, it will break. Your bones are the same.

Breaks have different names

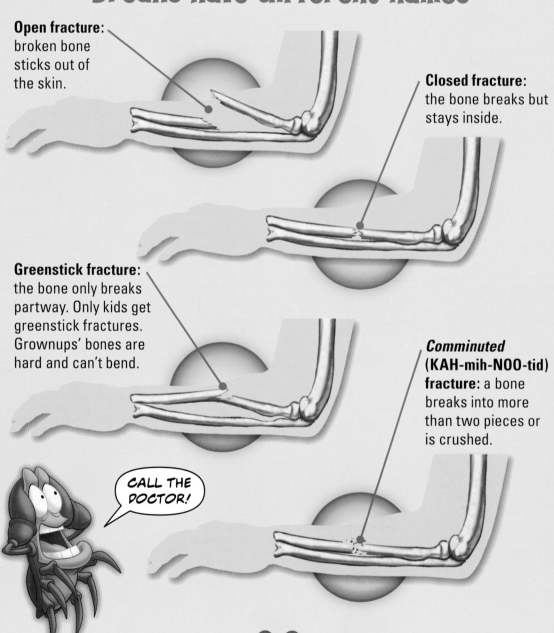

Open fracture: broken bone sticks out of the skin.

Closed fracture: the bone breaks but stays inside.

Greenstick fracture: the bone only breaks partway. Only kids get greenstick fractures. Grownups' bones are hard and can't bend.

Comminuted **(KAH-mih-NOO-tid) fracture:** a bone breaks into more than two pieces or is crushed.

CALL THE DOCTOR!

X-ray vision

Your bones are inside you, not outside where a doctor (or you) can see them. The doctor can't usually tell if you have a broken bone just by looking at you. To see your bones, the doctor takes an X-ray picture. X-rays are invisible rays that can pass through your body, but they don't go through hard parts like bone.

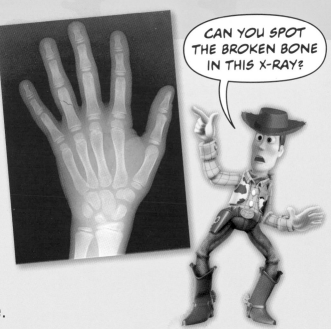

CAN YOU SPOT THE BROKEN BONE IN THIS X-RAY?

Fix it!

If you break a bone, the doctor makes sure the broken parts are lined up, then puts a cast around the part of your body with the broken bone. A cast is like a big, hard bandage. It holds the broken bone in place. Bones can then heal themselves!

Here's how it works:

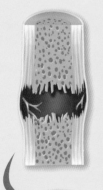

When a bone breaks, blood vessels break along with the bone. The space between the broken parts fills with blood.

A clump of cartilage forms and replaces the blood. The cartilage gets replaced by spongy bone.

When the bone starts getting used again, it puts stress on the spongy bone. The spongy bone solidifies into a hard, permanent patch.

Who Knew?

If a bone is badly broken into many pieces, doctors may use screws or metal plates to hold it together so it can heal.

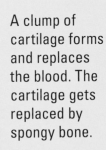

In and Out

You've scoped out your skeleton and the joints that let it move. But where do you get the energy to move in the first place? From the food you eat!

Starting gate

Digestion starts with your eyes and nose! When you see or smell something tasty, your mouth starts to water. Then you bite into the food. Your teeth chomp down and chew it into smaller pieces.

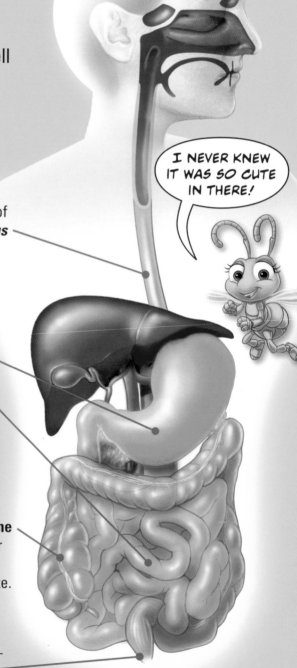

I NEVER KNEW IT WAS SO CUTE IN THERE!

When you swallow, the bits of food go down your **esophagus** (uh-SOFF-fuh-gus).

From there, the food goes to your **stomach** to be broken down even more.

More digestion happens in your **small intestine**. By the time the food gets to the end, all of the nutrients have gone into your body.

I STILL SAY IT'S YUCKY!

Your **large intestine** absorbs the water from what's left, leaving solid waste.

Finally, it's out through the exit— your **anus**.

14

Nutrition

The parts of your food that your body uses are called **nutrients**. Good nutrition means eating the right amount of these three main nutrients:

Carbohydrates, or starches and sugars, are broken down fast for energy. Vegetables, grains, and fruits have lots of carbohydrates.

Protein gives your body its building materials. Beans, peas, peanuts, fish, eggs, milk, and meat are foods that are high in protein.

Fat, in things like nuts and oils, is high in energy but it's harder for your body to break down. Your body needs fat, but a little goes a long way. Eating too much fat is unhealthy.

Toss the junk

If you eat a lot of junk food, you're getting a lot of empty calories, and you're not getting the vitamins, minerals, and other nutrients that your body needs.

IT'S ALL ABOUT BALANCE!

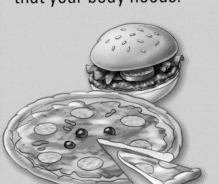

Fats and salts

Sugars

Chew on This

Your digestive system's job is to break down the food you eat. That's how it gets the nutrients out. It all starts in your mouth, where you use your teeth to chomp and chew and mash your food into tiny pieces.

Teeth

Take a look in the mirror and smile! See the different types of teeth in there? Each kind of tooth has a different job.

Incisors (in-SIZE-erz) are your front teeth. Feel the top edges of your incisors. They're shaped like little chisels. Your incisors are for biting and tearing off bite-size pieces of food.

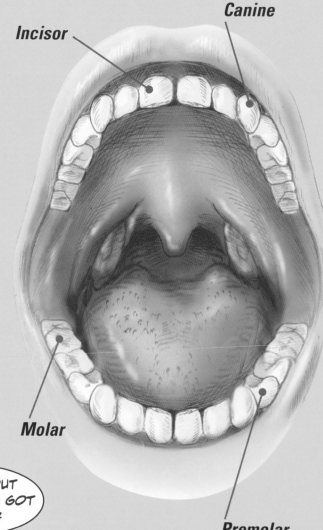

Incisor

Canine

Molar

Premolar

HOW ABOUT TUSKS? YOU GOT TUSKS?

Premolars (pre-MOLE-erz) and *molars* are your back teeth. Feel the tops of your back teeth with the tip of your tongue. They feel flat on top with little bumps and ridges. You use these to grind and mash food up into tiny bits.

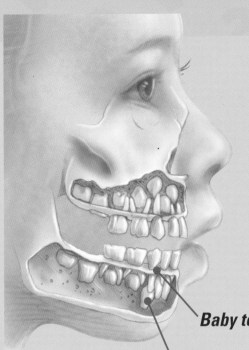

Baby teeth

Maybe you've lost a tooth or two. (Or three!) Your baby teeth are the ones you grew when you were still a baby. Your adult teeth grow in from above (in your upper jaw) or below (in your lower jaw) and push out the baby teeth. By the time you grow all of your adult teeth, you will have 32 of them.

Baby tooth

**Adult tooth
(still down in the jaw)**

SO LONG, PLAQUE!

Brush!

When you brush your teeth, you scrape away stuff on your teeth called *plaque* (PLAK). If you don't do this, cavities—little holes or cracks—can form in your teeth.

TASTE TEST

You can do an experiment to see how saliva (spit) changes starch to sugar. All you need is a soda cracker. (The experiment works best if the cracker is unsalted.) Just pop the cracker into your mouth and chew. Chew a lot. Chew it into tiny, tiny, tiny pieces. When the chewed-up cracker starts to taste sweet, you know that your saliva is changing the starch into sugar. (Oh yeah—you can swallow now.)

Mixing Bowl

Inside your stomach, your chewed food gets churned around and mixed with acid. Stomach acid is strong stuff! If you got pure stomach acid on your skin, it would hurt a lot. Your stomach has a lining of mucus on the inside to protect itself from the acid.

Digestive System

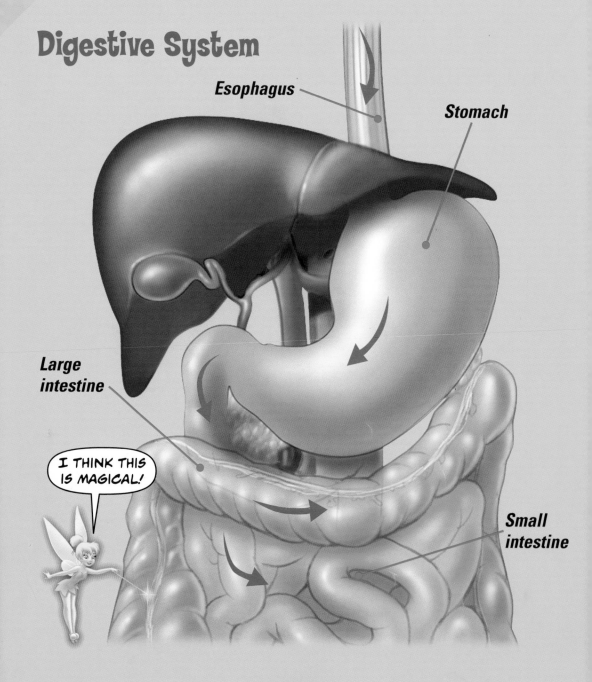

Esophagus

Stomach

Large intestine

I THINK THIS IS MAGICAL!

Small intestine

Growls and gurgles

Inside your stomach, chemicals work to digest what you eat. If there's not much food in there, the squeezing of your stomach muscles can slosh the liquids and gas around inside. That makes the gurgling noise of your stomach "growling."

GRRRR!

WHERE IS IT?

Put your hand on your stomach. Did you put your hand somewhere near your belly button? Your stomach is actually higher than that. It's up and to the left, partly under your bottom ribs.

The air down there

You often swallow a lot of air with your food. The air can rise up and get pushed back up your esophagus and out your mouth.

BURRRRP!

Some of the air and gas goes the other way. It travels all the way through your digestive system and comes out the other end.

EXCUSE ME.

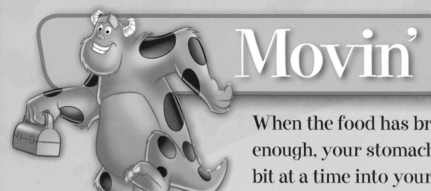

Movin' On

When the food has broken down enough, your stomach squirts it a little bit at a time into your small intestine.

The small intestine

The small intestine is anything but small! It's a long tube all coiled up inside you! It's called "small" because it's only about an inch wide, which is narrower than the large intestine.

The last steps of breaking down your food happen in your small intestine. Chemicals split apart the carbohydrates, fats, and proteins.

The inside of the intestine is lined with tiny *villi* (VILL-ee—a single one is called a villus). Each villus is shaped like a tiny finger, with walls only one cell thick. The nutrients are absorbed into the bloodstream through the villi.

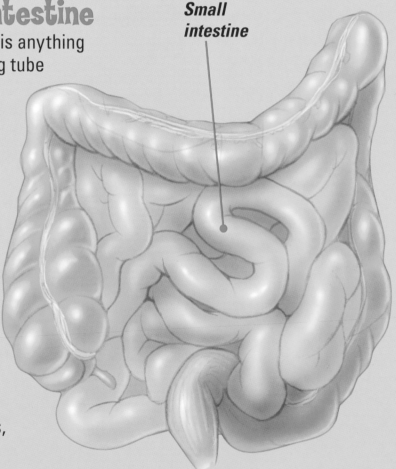

Small intestine

YOUR SMALL INTESTINE IS AS LONG AS A MINI-VAN!

End of the line

After your small intestine takes out all the nutrients it can, the water and undigested parts of the food pass into your large intestine. This intestine's job is to absorb water. Once most of the water has been taken out, all that's left of your food is the solid waste. This waste goes into the last part of your large intestine, called the **rectum**. It waits there, held in by the ring-shaped muscle at the end called the **anus**, until you go to the toilet and push it out.

WHAT REALLY HAPPENS IF YOU SWALLOW A PIECE OF GUM?

Gummy myth

Gum is mostly made of rubbery stuff that your body can't digest. That may have led to the myth that it takes seven years to digest gum. But you already know what happens to things that can't be digested. They come out with the rest of the waste. The whole trip from one end to the other takes about 20–40 hours.

HELPING OUT

Other organs help out with digestion, too.

- Your **liver** makes a green liquid called bile that mixes with the food in your small intestine. Bile helps break down fat.
- Your **gallbladder** stores bile until your body needs it.
- Your **pancreas** makes chemicals that break down all types of nutrients.

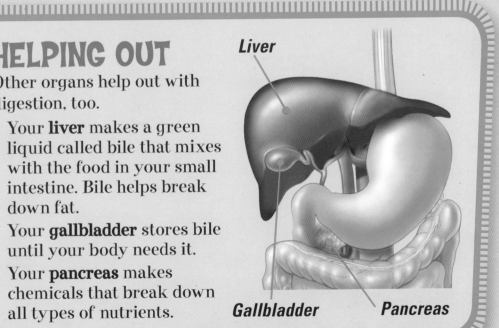

Liver

Gallbladder

Pancreas

Nervous System

If you want to ride your bike, sing a song, play a game, talk to a friend, or even take a nap, you'll need to use your brain! Almost anything you do with your body somehow involves your brain.

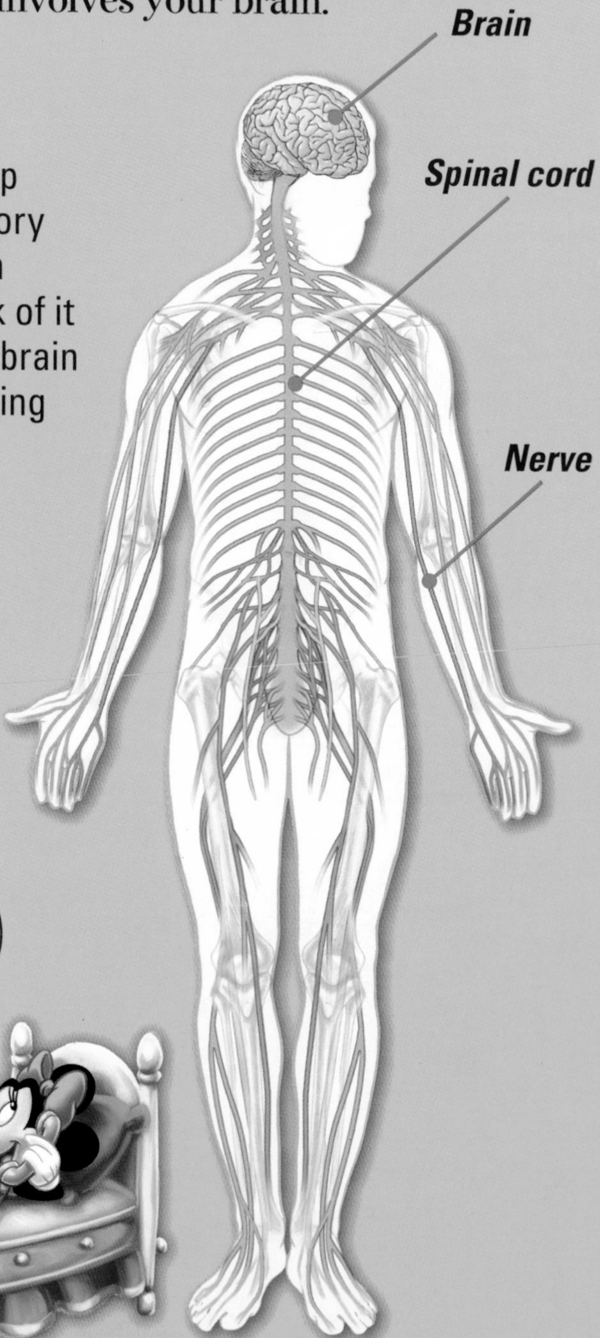

Message center

Your nervous system is made up of your brain, spinal cord, sensory organs, and the nerves that run through your whole body. Think of it as the wires that connect your brain and the rest of your body, relaying messages back and forth.

Brain

Spinal cord

Nerve

HELLO, BRAIN?

SORRY, ALL THE NERVES ARE BUSY. I'LL TELL THE BRAIN YOU CALLED!

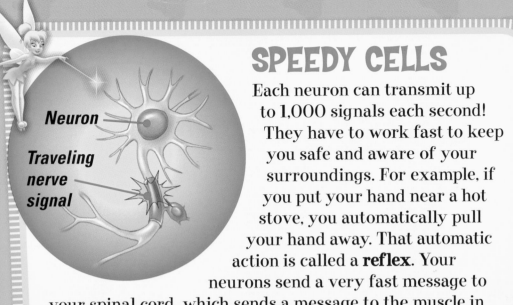

SPEEDY CELLS

Neuron

Traveling nerve signal

Each neuron can transmit up to 1,000 signals each second! They have to work fast to keep you safe and aware of your surroundings. For example, if you put your hand near a hot stove, you automatically pull your hand away. That automatic action is called a **reflex**. Your neurons send a very fast message to your spinal cord, which sends a message to the muscle in your arm, telling it to move away.

Your amazing brain

Your brain is divided into three parts:

- The **hindbrain** is the back, bottom part of your brain. Your hindbrain controls things like breathing and the speed of your heartbeat.

- The **midbrain** is above the hindbrain. It controls some reflexes and the movement of your body.

- The **forebrain** is the biggest part of your brain. It includes the *cerebrum* (suh-REE-brum). That's the folded, wrinkled part. The cerebrum is what lets you think, imagine, plan, learn, remember, speak, understand language, and do all the other amazing things you can do!

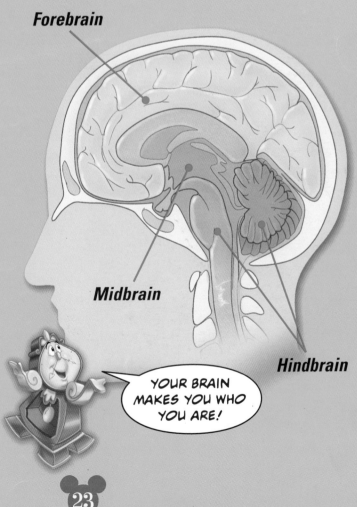

Forebrain

Midbrain

Hindbrain

YOUR BRAIN MAKES YOU WHO YOU ARE!

Senses

Your nervous system relays messages between different parts of your body. But how do your brain and body get information from the world around you? Through your senses!

You and eye

Vision is probably the sense that gives you the most information about the world around you. You're using it right now to read these words!

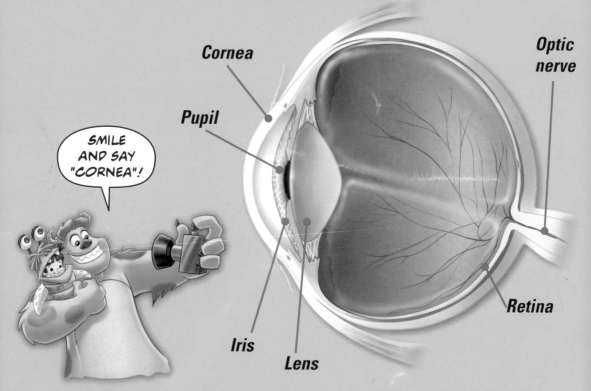

Cornea

Pupil

Optic nerve

SMILE AND SAY "CORNEA"!

Iris

Lens

Retina

- The front of each eye is protected by a clear cover—the **cornea** (COR-nee-uh).
- Light goes into your eye through a hole called the **pupil**. It's like a window into the eyeball. When the light is dim, your pupils get bigger. When the light is bright, your pupils get smaller.
- Next, the light passes through your eye's **lens**. The lens bends the light to cast an image on the back of the eyeball, called the **retina** (RET-in-uh).
- Special cells in the retina send signals to your brain through the **optic nerve**, which goes out the back of your eyeball. The brain uses the information from both of your eyes to create a 3-D picture of the things you see.

24

Eyes that need help

Your eye's lens focuses the image on the back of the retina. Many people have lenses that have trouble focusing, so they wear glasses or contact lenses to fix the problem.

SOME EYES NEED A HELPING HAND.

TRICK YOUR BRAIN

Your eyes just detect light, dark, and colors. They send that information to your brain. It's your brain that figures out what you are looking at. Sometimes your brain can be confused about what it sees. That's an optical illusion.

Do the sides of the square look curved? They are actually straight!

Do you see a triangle on top of the black circles? It's not really there!

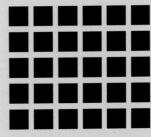

Do you see gray boxes where the lines cross? They're not really there!

NOBODY TRICKS MY BRAIN!

Hearing & Tasting

As you can see, those peepers of yours are pretty neat. Just as important, however, are your ears. And after that, it's on to the tongue!

Listen up!

You need your nervous system to hear and recognize all of the sounds around you. For example, if you snap your fingers, you create vibrations (or sound waves) in the air. That's sound! Sound vibrations travel through the air in waves.

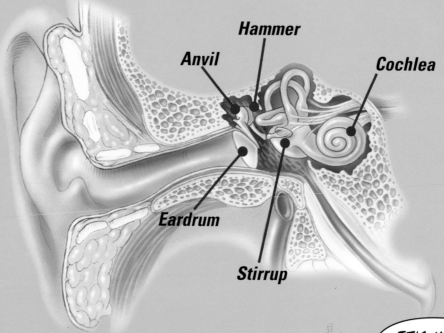

Hammer

Anvil

Cochlea

Eardrum

Stirrup

IT'S MUSIC TO MY EARS!

- Sound waves go into your ear canal, where they hit your **eardrum** and make it vibrate.
- Behind your eardrum are three small bones called the **hammer**, **anvil**, and **stirrup**. The vibrating eardrum makes these bones vibrate.
- Behind the bones is the *cochlea* (KO-klee-uh). It's like a tiny, rolled-up cone filled with fluid. The vibrating bones make the fluid inside the cochlea vibrate.
- Tiny hairs inside the cochlea detect the vibrations. The hairs are connected to nerves that send signals to your brain. You just heard a sound!

EARWAX

An inner part of your ear creates earwax: that shiny, sticky stuff that collects in your ears. Earwax is good for ears, so don't try to clean it out! It protects the inside of your ear from infections and traps dust, bugs, and other nasty stuff before it can enter your ear.

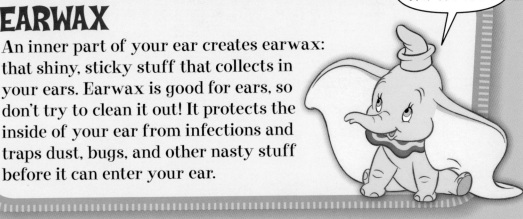

DID SOMEONE SAY EARWAX?

Tasty!

Stand in front of a mirror and stick out your tongue. See all those little bumps? Believe it or not, there are 10,000 of them! The bumps are covered with microscopic taste buds. Your taste buds can detect only five basic tastes: sweet, sour, salty, bitter, and a fifth taste that goes by the Japanese name *umami*. This taste was discovered by Japanese scientists. It's a taste found in mushrooms, aged cheese, and certain meats and soups.

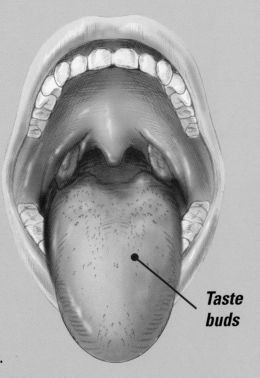

Taste buds

The nose knows

Smell is also a big part of your sense of taste. Hold your nose shut with one hand. Eat a little piece of apple. Now eat a little piece of raw onion. Do they taste different? When your nose can't smell, things seem to taste different from the way we expect them to.

Smell & Touch

Imagine if you could see and hear things, but couldn't feel or smell them. Wouldn't that be weird?

Hey! What smells?

It's your nose! Like taste, your sense of smell comes from molecule-detecting cells.

Receptors

It all starts when smell molecules go up your nose with the air you breathe. High up in the back of your nose are special sensing cells called *receptors* (re-SEP-ters). These smell cells have tiny hairs made of nerve fibers covered with mucus. When a smell cell detects the molecules of a smell in the mucus, it sends a message to your brain. Your brain identifies the smell.

Your sense of smell is 10,000 times more sensitive than your sense of taste. You can detect some smells from just a few molecules. You can recognize up to 10,000 different smells, from baking bread to stinky skunk.

Bark!

IT'S TRUE!

If you spread out the area of smell-sensing cells in your nose, it would be the size of a postage stamp. The smell-sensing cells of a bloodhound would cover an area the size of a handkerchief!

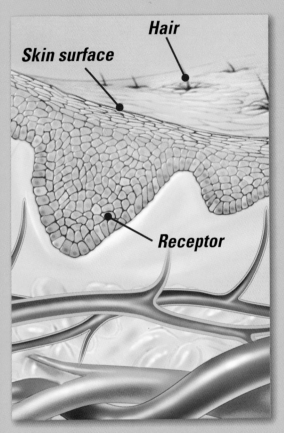

Hair

Skin surface

Receptor

Get the feeling

Your sense of touch comes from the receptors in your skin. There are many different kinds of skin receptors and they detect different things. There are receptors for light touch, heavier pressure, heat, cold, pain, and itching.

One finger or two?

Different areas of your skin have different numbers of receptors. Try this experiment:

1. Close your eyes. Get a friend to touch the palm of your hand lightly with either one or two fingers. Try to guess how many fingers are touching you.
2. Now have your friend do the same thing on your back. Is it easier or harder? How far apart do your friend's fingers have to be before you can feel two of them?

Sensitive areas, like your hands or lips, have lots of touch receptors. Less sensitive areas, like your back, have fewer receptors.

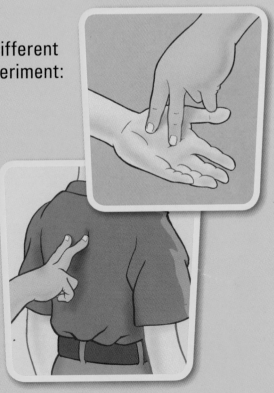

Muscles

Now you know that your brain is responsible for a lot of things your body does. That includes moving around! Even with a skeleton and a brain, you still need muscles to walk, run, skip, and dance. Without all those muscles, you'd be as stiff as a tree!

Many muscles

Your body has three different kinds of muscle tissue. **Smooth muscles** are made of layers or sheets of muscle cells. These are the muscles that work behind the scenes. They do the things your body does automatically, like pushing food through your intestines or changing the size of the pupils in your eyes.

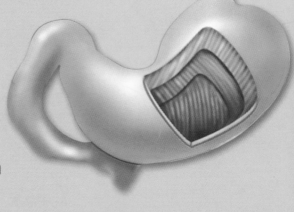

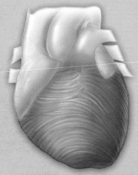

Cardiac (KAR-dee-ack) muscles are the muscles in your heart. These thick muscles squeeze to pump the blood through your body. They are the hardest-working muscles in your body. They keep pumping every minute of every day of your whole life.

Striated (STRI-ay-tid) muscles are also called skeletal muscles. These muscles attach to your bones. They are the type of muscle you feel and see when you bend your arm to "make a muscle."

THESE MUSCLES ONLY WORK WHEN YOU TELL THEM TO!

MUSCLE BASICS

Some muscles move inside where you can't see them. They do things like pumping your blood, pulling and pushing the air in and out of your lungs, and moving the food along in your digestive system. Other muscles move the parts you see, like your arms, legs, and head.

Who Knew?

Your body has about **650** muscles!

Fibers and strands

Muscles are made of bundles of long, threadlike cells called **muscle fibers**. Inside each cell are long, thin strands of muscle protein. Every muscle cell is controlled by a nerve cell. When the nerve cell sends a signal to contract, the muscle's protein strands pull and slide past each other. The cell gets shorter. When the nerve stops sending the signal, the protein strands let go. They slide past each other, and the cell can get longer.

Muscle fiber

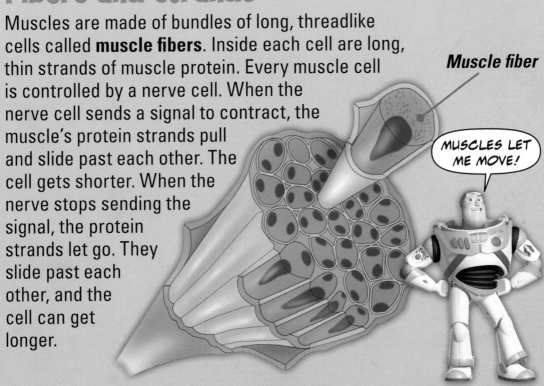

MUSCLES LET ME MOVE!

Push Me, Pull You

A muscle can only do work by contracting (getting shorter). When it relaxes, a muscle can only get longer if something pulls on it. So your muscles work in pairs. When one muscle contracts, the other one relaxes and gets longer.

Working together

You can see how it works by feeling the pair of muscles you use to bend and straighten your elbow.

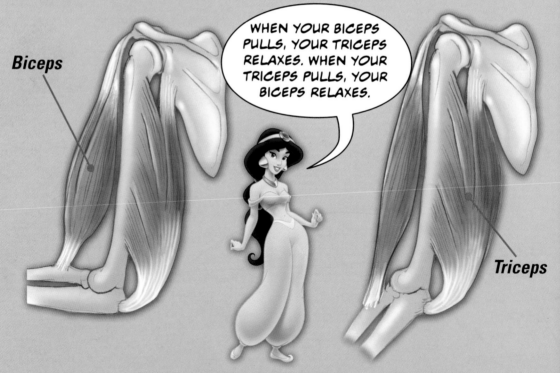

Biceps

Triceps

WHEN YOUR BICEPS PULLS, YOUR TRICEPS RELAXES. WHEN YOUR TRICEPS PULLS, YOUR BICEPS RELAXES.

- While holding one arm straight, rest your other hand on top of your upper arm. Now bend your arm. You will feel the muscle on top of your arm bunch up. That muscle is called the **biceps** (BI-seps). It pulls to bend your elbow.

- With your elbow still bent, feel the underside of your upper arm. That's the opposite side from your biceps. Now straighten your arm. The muscle you feel now is called the **triceps** (TRI-seps). It's the muscle that straightens your elbow.

Breathing muscles

Your **diaphragm** (DI-uh-fram) is a wide, flat sheet of muscle attached to your bottom ribs. You use your diaphragm to pull and push the air into and out of your lungs.

Your brain moves your diaphragm automatically, even when you're asleep. But you can also breathe in and out or hold your breath when you want to.

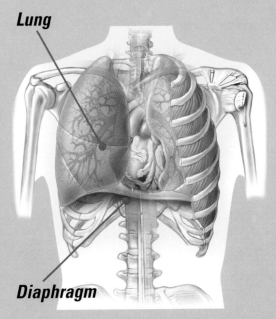

Lung

Diaphragm

Who Knew?

Usually your diaphragm moves smoothly, but sometimes it gets irritated and moves with a quick jerk. That's a hiccup.

Use 'em or lose 'em

Your muscles need to work to stay strong. The harder you work your muscles, the stronger they get. Muscles get stronger by making more muscle fibers. So when a muscle gets stronger, it gets bigger. That's why athletes, like weight lifters, bicyclists, and speed skaters, have such big muscles. If you don't use your muscles, they get weaker. Weaker muscles actually break down and get rid of the fibers they aren't using.

WORKING OUT IS HARD WORK!

More on Muscles

Your 650 muscles have more surprises in store for you!

Muscle hall of fame

Biggest: the *gluteus maximus* (GLOO-tee-us MACK-sih-mus), otherwise known as your rear end! You have two of these muscles, one for each leg. The gluteus maximus is the muscle you use to stand up from a squatting position.

Smallest: the *stapedius* (stuh-PEE-dee-us). It is connected to your smallest bone, the stirrup bone in your inner ear.

Strongest: the biting muscles in your jaw.

Longest: the *sartorius* (sahr-TOR-ee-us). This muscle starts at your lower spine, runs across the top of your thigh, and attaches to your shinbone. You use your sartorius muscles to rotate your thigh outward and bend your lower leg up and in.

WHERE ARE THE VINE-SWINGING MUSCLES?

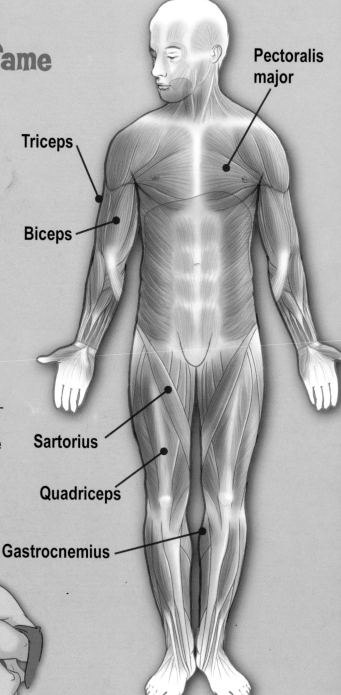

Pectoralis major

Triceps

Biceps

Sartorius

Quadriceps

Gastrocnemius

Holding steady

It takes muscle power just to sit up straight or stand in one place. If these muscles let go, you would fall down in a heap on the ground! That's what happens when people get knocked out or become unconscious. The brain stops sending signals to the skeletal muscles, and all the muscles relax. And then?
THUD!

Growing pains

Some kids find that the muscles in their legs hurt once in a while. If this happens to you, you might have growing pains! But don't worry— they are just a sign that your body is growing. Sooner or later, you won't get them anymore.

FLOATERS

1. Stand in a doorway with your arms down at your sides. Put the backs of your hands on the door frame and push outward, hard. Stay that way while you count slowly to 30.
2. Now step out of the doorway. Relax your arms and breathe in.

Your muscles are always working to hold your body in its normal position. When you were pushing out on the door frame, your arm muscles got used to that position as "normal." When you stepped out, your **deltoid** (shoulder) muscles were still pushing out, but there was no doorway. Instead they pushed your arms upward!

TAKE IT FROM A NATURAL FLOATER: THIS IS FUN!

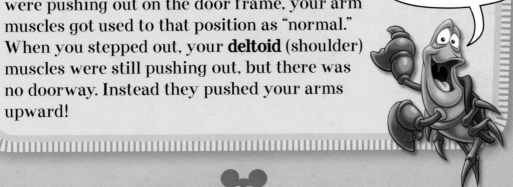

Cardiopulmonary

What's the hardest-working muscle in your body? It's your heart! Even when you're lazing around on the couch, your heart is pumping away. It takes *less than a minute* to pump blood to *every one of your cells.* That's amazing, considering there are billions of cells in your body!

The pump

Your heart is the pump that keeps your body going. It pumps the blood that carries oxygen to every cell in your body. About the size of your fist, your heart pumps about 51 million gallons of blood during an average lifetime!

Your heart is really two pumps in one. One side pumps blood from your lungs into the rest of your body. The other side pumps blood from your body into your lungs.

Your heart beats at different speeds depending on how much work your body is doing.

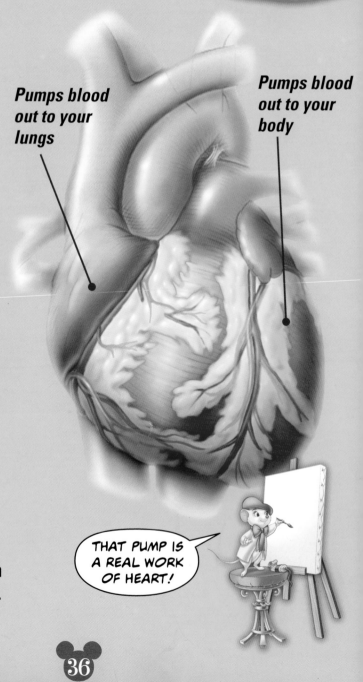

Pumps blood out to your lungs

Pumps blood out to your body

THAT PUMP IS A REAL WORK OF HEART!

Lub-dup! Lub-dup!

A heart makes a double-beat that sounds like "lub-dup." You can make your own stethoscope to listen to a friend's heartbeat. Just cut the bottom off a paper cup. Put the wide end of the cup on the middle of your friend's chest. Put your ear to the narrow end. Do you hear the heartbeat?

Who Knew?

When you "put your hand on your heart," you probably put it on the left side of your chest, but your heart is really almost right in the middle.

FAST BEATS

To do this experiment, you will need a watch with a second hand, a stopwatch, or a digital watch that shows seconds, and a pencil and paper.

1. Find your wrist pulse. Hold your hand out with the palm facing you. Now tilt the hand (not the arm) away from you. Using two fingers, press gently just below your wrist on the thumb side. You may have to slide your fingers up or down to find the pulse.
2. Count the number of pulses in one minute (60 seconds).
3. Now jump up and down for one minute.
4. Okay, stop! Measure your pulse again. Write the number down. Which number is higher? Why do you think that is?

Blood

Oxygen is a gas in the air you breathe. Your blood's main job is to carry oxygen and nutrients to all the cells in your body, and to carry away waste from the cells.

Blood cells

Blood is a liquid with cells floating in it.

Red blood cells look like little flying saucers. They carry the oxygen.

YOU CAN'T SEE THESE DIFFERENT CELLS, BUT THEY ARE THERE.

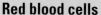

White blood cells look like blobs. Their job is to attack and fight germs in your body. There are many more red blood cells than white ones.

Platelets (PLATE-lets) help your blood to clot and form a scab when you get a cut or scrape.

Plasma is the liquid part of blood. The blood cells float in the plasma. In the plasma there are also nutrients and other chemicals used by your body.

THE AVERAGE GROWNUP HAS AROUND 10 PINTS OF BLOOD. THAT'S ALMOST A GALLON AND A HALF!

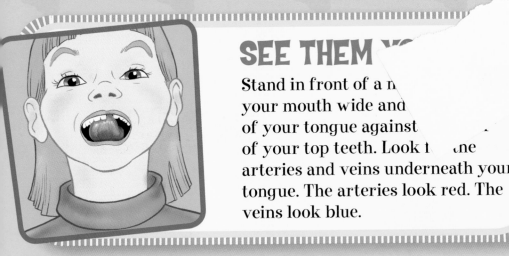

SEE THEM Y...

Stand in front of a m...
your mouth wide and...
of your tongue against...
of your top teeth. Look f... the
arteries and veins underneath your
tongue. The arteries look red. The
veins look blue.

On the move

Blood moves around your body in blood vessels of the *circulatory* (SIR-cue-luh-tor-ee) system. There are three types of blood vessels:

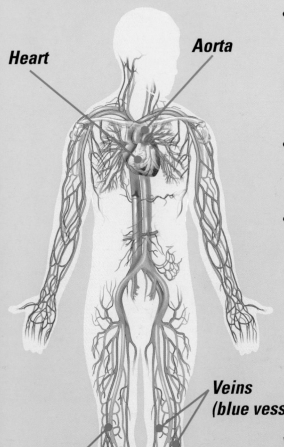

Heart

Aorta

Veins (blue vessels)

Arteries (red vessels)

- **Arteries** are the blood vessels leading away from the heart. In grown-ups, the largest artery in the body (the aorta) is almost as thick as a garden hose! The aorta is located in your heart.

- **Veins** are the blood vessels leading toward your heart. Almost all veins carry blood with the oxygen used up.

- **Capillaries** are very, very tiny blood vessels. They are only as wide as one red blood cell, so the blood cells have to travel through them single file! Capillaries carry the blood cells right up to your body's tissue cells.

Bug blood

Earthworms and insects have green blood! Starfish have yellowish blood. And crabs and lobsters have blue blood!

Lungs

Your lungs are your body's air-exchange system. You use them to bring in oxygen and get rid of carbon dioxide.

In and out

When you breathe, you fill your lungs up with air and push it back out again. You move the air in and out using your **diaphragm**.

- The diaphragm pulls downward to suck air into your lungs. The air comes in through your nose and mouth, down your windpipe.

- The diaphragm moves upward. The air goes out again.

BREATHE IN! BREATHE OUT! NOW YOU'VE GOT IT!

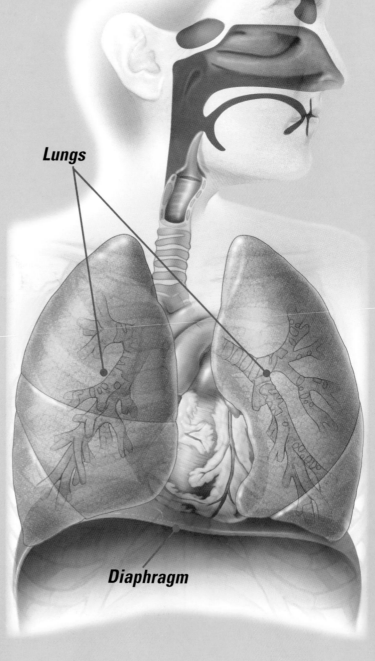

Lungs

Diaphragm

Branches

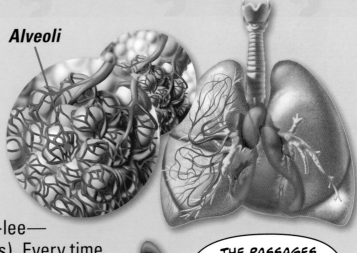

Alveoli

Your lungs aren't just like big balloons full of air. Each lung has many branching air passages. At the ends of the smallest passages are groups of tiny air sacs called *alveoli* (al-VEE-uh-lee—a single one is an alveolus). Every time you breathe in oxygen, it passes into the alveoli, then through capillaries, and onward throughout your body.

THE PASSAGES GET SMALLER AND SMALLER AS THEY BRANCH OUT!

Who Knew?

If you stretched out all of the alveoli you have in your lungs, they could cover a tennis court!

TIME YOUR BREATHS

You'll need a watch or clock with a second hand, a stopwatch, or a digital watch that shows the seconds, and a pencil and paper.

1. Count the number of breaths you take in 60 seconds. One breath is a full cycle, in and out. Write the number on the paper.
2. Now jump up and down for a full minute.
3. Time your breaths again. Write down that number. Which breathing rate is faster?

When you exercise, your cells use more oxygen. To get more oxygen, you breathe faster and more deeply.

Good In, Bad Out

When you exhale, you breathe out carbon dioxide, a waste product that your cells want to get rid of. So, your lungs not only help to bring oxygen into your body, but they also remove the carbon dioxide from your blood.

The gas exchange

The alveoli in your lungs are surrounded by capillaries. The walls of the alveoli are only one cell thick. So are the walls of the capillaries. The place where the alveoli and capillaries meet is like a pickup and drop-off area for oxygen and carbon dioxide.

Pick up. . .
Oxygen from the air in the alveoli passes into the capillaries, where it gets picked up by the red blood cells. Your heart pumps the blood—the cells carry the oxygen to the other parts of your body.

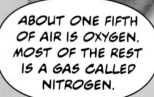
ABOUT ONE FIFTH OF AIR IS OXYGEN. MOST OF THE REST IS A GAS CALLED NITROGEN.

And drop off. . .
Blood coming from the body contains carbon dioxide, picked up from the cells. Your heart pumps the blood to the alveoli. The carbon dioxide in the blood passes into the alveoli, where it gets pushed out of the lungs when you breathe out.

Tiny brooms

Particles of dust, dirt, smoke, and other pollution in the air get cleaned out of your lungs by tiny hairs called *cilia* (SIL-ee-uh). They sweep up the dirt and pass it up your windpipe in mucus. When you cough, you're pushing out the dirt and mucus.

Cilia

Keep it clean

Your lungs get rid of the carbon dioxide in your blood, but there are other waste chemicals, such as ammonia, that need to be taken out, too. That happens in your kidneys. These two organs are near the middle of your back, on either side of your backbone. Your kidneys have several jobs, but one of the main ones is to filter waste chemicals out of your blood. The kidneys mix the wastes with water to create urine.

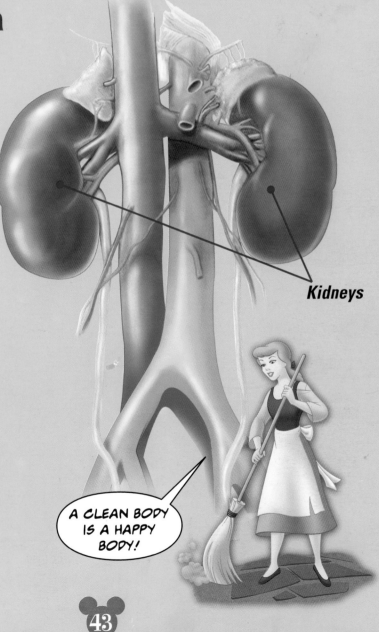

Kidneys

A CLEAN BODY IS A HAPPY BODY!

Wrap It Up

One of the most important parts of your body is on your inside *and* your outside. It's the biggest organ in your body, too! What is it? It's not your stomach, your lungs, or your heart. It's your skin! It covers you from head to toe. Your skin doesn't just hold your bones, muscles, and organs inside. It protects you from germs and weather, and keeps your body at the right temperature.

Your epidermis is showing!

Your skin has two main layers:

- The outer layer is called the *epidermis* (ep-ih-DER-miss). The inside cells of this layer are always dividing, making new cells. The outside cells are hard and tough, but eventually they die and flake off. New cells are always moving outward to replace them.

- The inner layer is called the **dermis**. This is where the blood vessels, nerves, and sweat glands are. It's also the place your hair grows from.

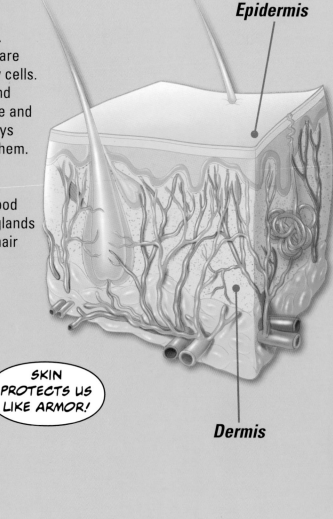

Epidermis

Dermis

SKIN PROTECTS US LIKE ARMOR!

HE SHOULD HAVE WISHED FOR AN AIR CONDITIONER!

Keep your cool

Your skin makes sweat to cool off your body. When you're hot, sweat from your sweat glands comes out of tiny holes on the surface of your skin, called pores. From there, the sweat evaporates, carrying the heat away into the air. Your body loses heat. That cools you down to help keep you comfortable.

Are you thick-skinned?

On most of your body, your skin is about two millimeters thick (less than one tenth of an inch). The skin of your eyelids is only one half of a millimeter thick. Your thickest skin is on the soles of your feet, where it is about six millimeters (one quarter inch) thick.

VERY HAIRY IS VERY SCARY!

HAIRY FACTS

- There's hair growing on almost every square inch of your skin! There are only a few places where your skin has no hair, including the soles of your feet, the palms of your hands, your lips, and your belly button.
- Each hair grows out of a hair **follicle** (FOL-ih-kul) in your skin. The hair grows from the root.
- Except for the growing part of the root, your hair isn't alive. That's why it doesn't hurt when you get a haircut.
- The 100,000 hairs on your head grow about half an inch every month. About 50 or more of them fall out every day, but they get replaced by new hairs.

THE QUIZ

The answers to all these questions are somewhere in this book. How many do you remember? How many can you find?

? What's the smallest bone in your body? *(page 7)*

? What is the name for the places where the bones in your head meet? *(page 9)*

? How many taste buds are on your tongue? *(page 27)*

? What is the name for your chisel-shaped front teeth? *(page 16)*

? Where could you find cardiac muscles? *(page 30)*

? What is the name for the clear covering at the front of your eyes? *(page 24)*

? What are the three main nutrients we get from food? *(page 15)*

? What is the name of the stuff you scrape off your teeth with a toothbrush? *(page 17)*

? How old will you be when your bones stop growing? *(page 8)*

? What is the longest muscle in your body? *(page 34)*

? What part of the brain lets you think, learn, and remember? *(page 23)*

PUT YOURSELF TO THE TEST!

? What is the hardest substance in your body? *(page 7)*

? How many muscles does it take to speak? *(page 4)*

? What kind of joint is your elbow? *(page 10)*

? What is the name for a bone fracture where the bone only breaks partway? *(page 12)*

? What color is a lobster's blood? *(page 39)*

? How many bones are in your skull? *(page 6)*

? How many hairs fall out of your head every day? *(page 45)*

? What is the name for special sensing cells? *(page 28)*

? What is the name of the big sheet of muscle below your ribs? *(page 33)*

? Which organ makes bile? *(page 21)*

GET THAT BRAIN WORKING!

? What is another name for a nerve cell? *(page 4)*

? What kind of blood cell carries oxygen? *(page 38)*